EXPLORING THE SEA

Scientific Consultant:
Steven L. Bailey
Curator of Fishes
New England Aquarium

Photo Credits:
Wayne & Karen Brown—Endpages; Pages 8, 15, 20, 26
Michael J. Giudice—Pages 7, 27
Francois Gohier—Cover; Pages 15, 24, 25, 27
Tom & Pat Leeson—Page 25
NASA—Page 6
Norbert Wu—Pages 7, 19, 22, 23, 24, 28, 29
Al Giddings Images Inc.—Pages 21, 22
Vince Cavataio/Allsport USA—Page 12
AP/Wide World Photo—Pages 13, 20
AP/National Maritime Museum—Page 21
The Bettman Archive—Page 10
Gamma Liaison—Cover; Page 11
Kaku Kurita/Gamma Liaison—Page 29
Pierre Perrin/Gamma Liaison—Page 27
Gamma Tokyo—Page 28
The Granger Collection—Pages 10, 11, 20, 24, 29
Doug Perrine: Innerspace Vision—Page 8
J.L. Dugast/Liaison International—Page 6
Lyle Leduc/Liaison International—Page 10
Carl Schneider/Liaison International—Page 13
Hamblin/Liaison International—Page 14
Paul Kennedy/Liaison International—Page 26
Paul Souders/Liaison International—Page 26
Nance Trueworthy/Liaison International—Page 14
Brandon Cole/Mo Young Productions—Page 14
Mark Conlin/ Mo Young Productions—Pages 19, 29
Darodents/Pacific Stock—Page 6
Reggie David/Pacific Stock—Page 18, 19
Sharon Green/Pacific Stock—Page 7
William Bacon/ Photo Researchers—Page 25
Chesher/Photo Researchers—Page 21
David Hardy/Science Photo Library/Photo Researchers Inc.—Pages 13, 19
NASA/Science Photo Library/Photo Researchers Inc.—Page 12
Hal Beral/Visuals Unlimited—Page 9
David B. Fleetham/Visuals Unlimited—Page 18
Will Troyer/Visuals Unlimited—Page 9
Kevin Deacon/Waterhouse Stock Photography—Page 26
Stephen Fink/Waterhouse Stock Photography—Cover; Pages 16, 17
Dave Fleetham/Waterhouse Stock Photography—Cover
Marty Snyderman/Waterhouse Stock Photography—Pages 8, 23
James Watt/Waterhouse Stock Photography—Page 9

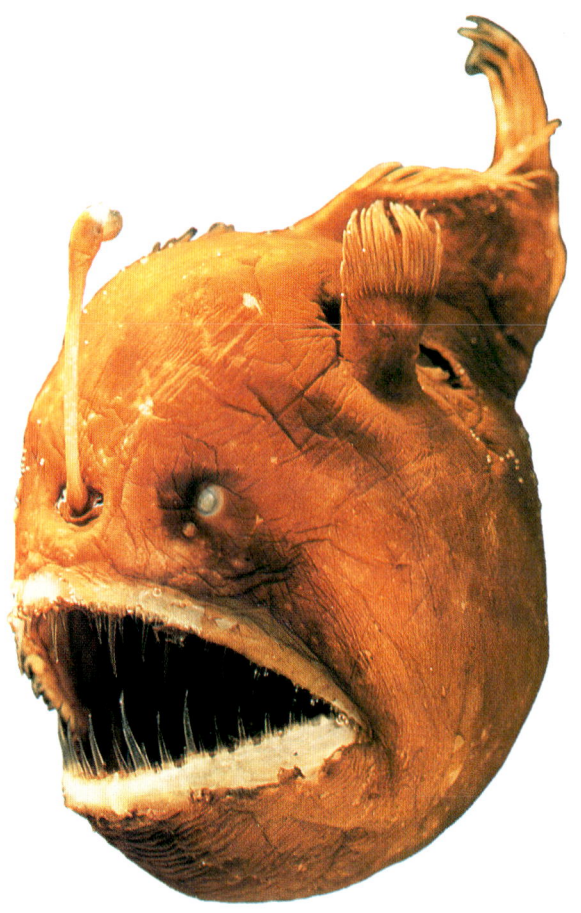

Illustrations
Howard S. Friedman—Pages 9, 12, 18, 19

Copyright © 1998
Kidsbooks, Inc.
3535 West Peterson Ave.
Chicago, IL 60659

All rights reserved including the right of
reproduction in whole or in part in any form.

Manufactured in the United States of America

Visit us at www.kidsbooks.com
Volume discounts available for group purchases.

EYES ON ADVENTURE

Exploring

THE SEA

Written by
Celia Bland

kidsbooks
Incorporated

WATERY PLANET

If you were to rocket into space and look back at Earth, you would see a big blue planet. It looks blue because most of Earth's surface is covered with water. The sea, in fact, has one thousand times more room for living creatures than air and land combined.

GREAT OCEANS

The Pacific, Atlantic, Indian, and Arctic oceans cover over two-thirds of our planet. That's more than 300 million square miles! The Pacific is by far the deepest and the largest ocean. It covers more than one-third of the globe.

PASS THE SALT

What's the saltiest ocean? The Atlantic. Rocks make the water salty. Waves erode the rocks, which contain salt that dissolves in the water. Some sea animals can't live in very salty waters. Others, like clams and oysters, use the calcium found in sea salt to build their shells.

Salt literally drips from rocks near the Red Sea—one of the saltiest seas on the planet.

WATER RE-CYCLE

How do Earth's oceans remain so full of water? The answer is in the water cycle. When water evaporates from the sea it becomes rain clouds. When the rain falls onto land, it drains into rivers. Then the rivers take water back to the sea.

DOWN UNDER
Home to many fascinating creatures, and burial ground for countless shipwrecks, the sea is a gold mine to adventurous types. Divers investigate animal behavior, recover ancient wrecks, and discover new marine species.

A diver observes a giant seajelly in action. ▼

FIRE AND ICE
Close to the equator, where the climate is hotter than anywhere else on Earth, sea water is warm. Farthest from the equator lies the Arctic Ocean, home of glaciers, icebergs, and animals specially adapted to wintry weather. Here the sea is icy cold.

SEA STUDY
Scientists who study the oceans are called *marine biologists*. They try to figure out how and why the oceans change, and why certain sea creatures and plants live in one place and not in another.

NOW AND THEN
About 6,000 years ago the ancient Egyptians invented sails. Over the centuries, sailing became a vital means of transportation and industry. Today, sailing is a pastime and a sport. One of the world's most famous ocean sailing races is the America's Cup.

WHO LIVES HERE?

Try to imagine all the living things in the world—more than 10 million species of animals, plants, fungi, bacteria, and other types of creatures! Does it come as a surprise that only 20 percent live on land? The remaining 80 percent are found in the sea.

◀ The octopus was once thought to be a monster.

▲ The hermit crab scuttles across the sea floor.

SINK OR SWIM
When you think of water, you probably think of swimming. But not all sea creatures spend their life swishing their fins like fishes or whales do. Many plants and animals live on the sea floor. Tiny plants and animals known as plankton simply float on the ocean currents.

◀ The wide-mouthed manta ray easily gobbles up some plankton.

UP FOR AIR
Sea mammals, such as whales and dolphins, cannot spend all of their time under the water's surface like fish. They must come up to breathe, as people must do when in the water. However, sea mammals can dive for long periods of time. The sperm whale can stay underwater for more than an hour, holding its breath while hunting giant squid.

HOMEBODIES

Many sea creatures stay in one area of an ocean their whole life. Certain animals, such as the giant whale shark at right, roam the waters for food.

Hatched in rivers, salmon live in the ocean during their adult life. But when it comes time to spawn, or produce young, salmon leave the ocean and swim back to the river where they were born.

STAYING ALIVE

As a lower link on the food chain, small fish have developed a great defense—swimming in schools. Because the fish swim together, darting left and right, predators have a hard time picking out a single fish to catch.

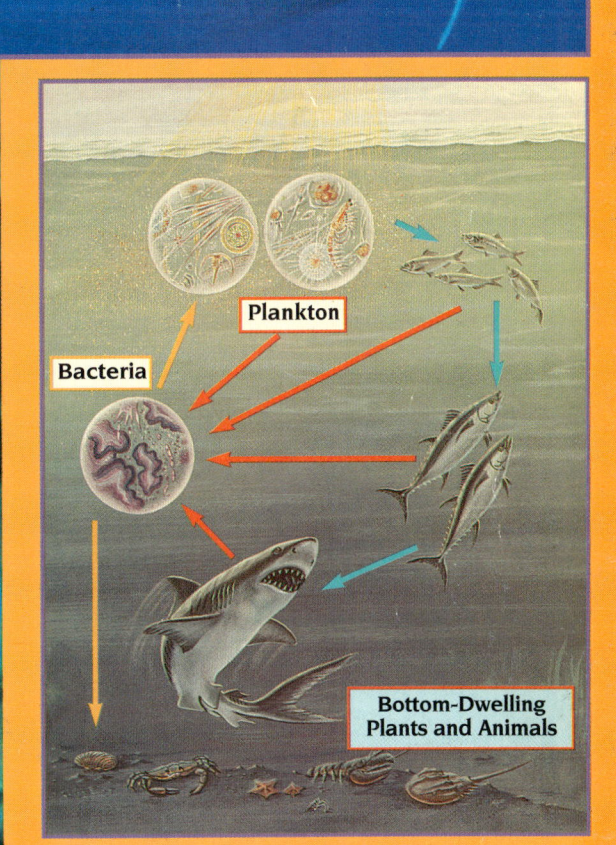

THE FOOD CHAIN

Like any chain, a food chain is made of links—living creatures eating other living things. It all starts with bacteria, which is partly dependent on the decomposition of dead animals. Bacteria provides nutrients to plankton and other sea life. Then plankton are eaten by small animals who are in turn eaten by larger animals.

ALL ABOARD

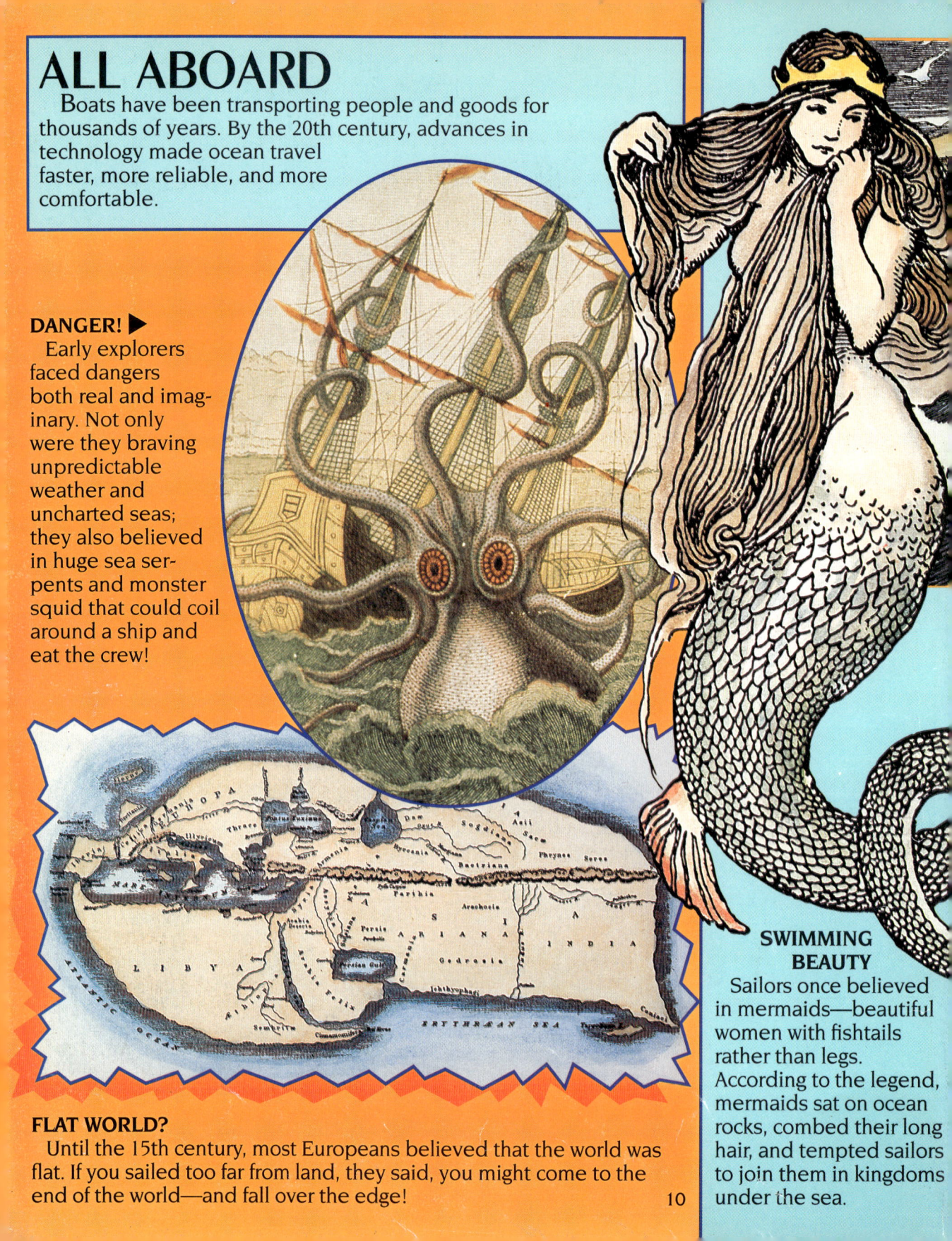

Boats have been transporting people and goods for thousands of years. By the 20th century, advances in technology made ocean travel faster, more reliable, and more comfortable.

DANGER! ▶

Early explorers faced dangers both real and imaginary. Not only were they braving unpredictable weather and uncharted seas; they also believed in huge sea serpents and monster squid that could coil around a ship and eat the crew!

FLAT WORLD?

Until the 15th century, most Europeans believed that the world was flat. If you sailed too far from land, they said, you might come to the end of the world—and fall over the edge!

SWIMMING BEAUTY

Sailors once believed in mermaids—beautiful women with fishtails rather than legs. According to the legend, mermaids sat on ocean rocks, combed their long hair, and tempted sailors to join them in kingdoms under the sea.

VIKING MIGHT

Some of the earliest explorers were Vikings. Eric the Red sailed all the way from Norway to Greenland around A.D. 980. His son Leif Eriksson sailed even further—from Iceland to the east coast of Canada! This was around A.D. 1000—nearly 500 years before Columbus.

AROUND THE WORLD

Ferdinand Magellan is credited with circling the globe and proving the world was round, but it was his navigator, Juan Sebastian del Cano, who captained the ship that finished the voyage in 1522. Magellan was killed in a battle the year before.

DEEP-SEA VESSELS

For centuries, people traveled *on* the water. Not until the 18th century, with the invention of the submarine, was *underwater* travel made possible. From the submarine came submersibles—small submarines used for marine research, archaeological expeditions, and pleasure rides.

▼ Today, ocean travel can be a vacation. Cruise ships offer luxury and spectacular views.

ON THE MOVE

The ocean stays in motion. Even when the water seems calm—waves, tides, and currents keep the waters flowing.

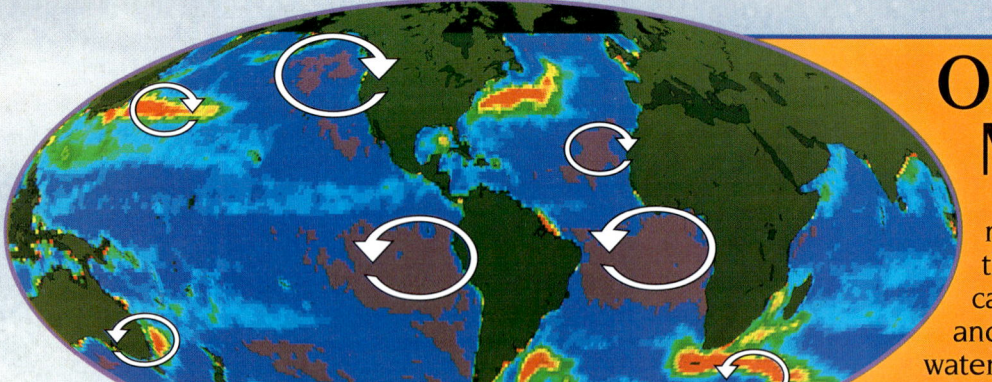

CURRENT EVENTS

Not until the 20th century did people begin to understand the ocean's currents. In the Northern Hemisphere, currents sweep clockwise from the equator to the Arctic. In the Southern Hemisphere, they travel counterclockwise.

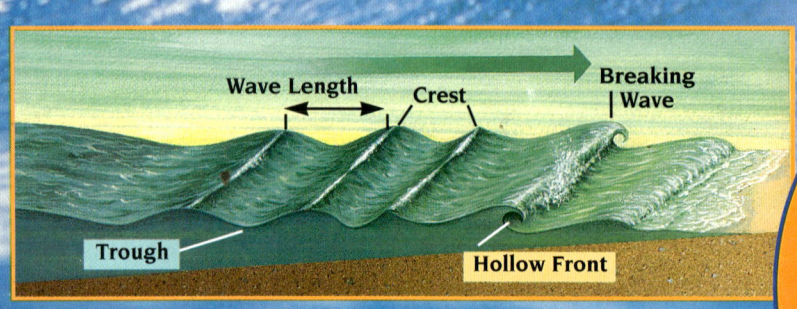

TIME AND TIDE

Tides are caused by the sun and moon's gravitational pull. As Earth turns on its axis, one side of the planet comes closer to the moon. Pulled by the moon's gravity, the ocean may rise toward the moon and become deeper, which is called high tide. On the opposite side of Earth, the ocean may be more shallow—called low tide.

MAKING WAVES

How do waves form? Wind blowing along the sea's surface drags the top of the water and creates waves. As a wave moves closer to shore, the water below the wave gets shallow. This makes the wave taller until it moves into even shallower water, topples over, and breaks.

BIG BOARDS
Surfing has been an exciting sport since A.D. 400, when it was invented by the Polynesians in the Pacific. The first Westerner to witness surfing was Captain James Cook at Kealakekua Bay, Hawaii, in 1778. The surfboards, shaped from trees, were 20 feet long and weighed as much as 200 pounds!

◀ KILLER WAVES
Tidal waves, or *tsunamis*, don't actually have any connection to tides. They are caused by seismic activity on the sea floor. Tsunamis can travel hundreds of miles, growing sometimes hundreds of feet high, and reaching speeds of 500 mph.

WATER SPORTS ▼
A combination surfboard and sail, "sailboards" are speedy and acrobatic. Some windsurfers have been clocked at 50 mph, faster than any other sailing craft!

▲ KON-TIKI EXPEDITION
In 1947, explorer Thor Heyerdahl built a simple wooden ship similar to the one used by early peoples. He wanted to test whether Indians from South America could have sailed the seas to settle Polynesia. It took him 101 days to sail from Peru to Polynesia, which proved the early expedition possible.

MEETING THE LAND

The coast can be a loud and dramatic place, where rocky cliffs are hit hard by breaking waves. Coastal creatures have to cling to rocks to keep from washing out to sea. Barnacles, snails, and bivalves—such as mussels and some types of scallops—attach themselves to coastal rocks with a sticky secretion produced in their body.

PUDDLE DWELLERS
When the tide goes out, puddles of water are left on shore. Seastars, seaweed, periwinkle snails, crabs, and sea urchins make their homes in these tide pools.

SHELL TREASURES
Have you ever found a queen conch shell on the beach? How about a tiger cowry or a rosy harp? These beached shells were once the homes of soft-bodied animals known as shellfish, or mollusks. Empty shells are popular collectible items for beachcombers.

WEED FOREST
Kelp, or seaweed, grows in "forests," taking root on hard sea floors near shore. The fastest-growing plant, kelp can gain as much as three feet a day, and reach lengths of 213 feet! Kelp floats upward instead of sinking because its stems are surrounded by gas-filled bulbs.

OTTER HANG-OUT
Kelp forests serve many purposes for the sea otter. During the day, otters eat the abalone, crabs, and urchins that feed there. At night they wrap themselves in the kelp fronds, and the waves rock them to sleep.

CITY OF CORAL

One of the most colorful and populated areas of the sea is in and around a coral reef. The bright colors of the coral are caused by algae that live inside. Outside the coral, schools of tropical fishes dazzle the eye with an equally fascinating display of color.

NO TOUCHING!

One of the best ways to appreciate the living treasures of a coral reef is to see it up close. Snorkelers and scuba divers flock to reefs. But they are careful never to touch the coral because that can damage or kill it.

IT'S ALIVE!

Coral may look and feel stony, but it is not rock. It's the skeleton of a living animal called a polyp. The polyps grow a skeleton on the outside to protect and support their soft bodies. Because the reef-building corals cannot live in water colder than 64°F, they are found only in warmer waters.

FEEDING TIME

All animals have to eat, including coral polyps. How do they do it when they are attached to the ocean floor? Coral polyps actually have tiny arms that catch plankton and pass it into their mouth.

THREE REEFS

Reefs grow in different ways. A *fringing* reef is attached to the shore. An *atoll*, like the one shown at left, is a ring of coral formed around a sunken volcano. A *barrier* reef has a channel of water between it and the shore. Australia's Great Barrier Reef is a whopping 1,250 miles long. That makes it the biggest structure ever built by animals.

COAT OF ARMOR

You can often count on finding clownfish with the grasslike sea anemone. Unlike most sea life, the clownfish is safe from the anemone's stinging cells because of a thick, slimy mucous on its body.

Brain coral

KNOW YOUR CORAL

Stony corals, or hard corals, such as brain corals, form reefs. Gorgonians, or soft corals, such as sea fans, grow on the sea floor and on reefs, and look a lot like ferns or bushes.

CORAL POPULATION

Every coral reef has a population consisting of thousands of different animals that live and thrive there—including shellfish, moray eels, sea horses, and sharks.

SEASCAPES

Under the ocean there is a landscape almost as varied as the one above water—with gigantic canyons, plains, mountains, and caves.

AGE-OLD SEA
When plants and animals die in the ocean, their remains drift down to a sea bottom. Drilling deep with special tools, scientists take samples of this sediment. The samples provide a rich history of millions of years of ocean life.

At the center of the Atlantic Ocean is a mountainous ridge surrounded by plains and valleys.

North America

South America

Europe

Africa

UNDERWATER MOUNTAIN
Most islands are really the peaks of underwater volcanic mountains. Movements in the earth's crust can produce heat and pressure inside an underwater volcano. The pressure eventually causes the volcano to "blow its top." Lava, dust, and rocks flow out, covering the volcano layer by layer, until it breaks the ocean surface and makes an island.

Sea Level

Crust

Mantle

ECHO MEASURE
To measure the ocean's depth, scientists transmit sonar (sound) pulses toward the sea floor, then listen for the echo. The longer it takes to hear the echo, the deeper the water.

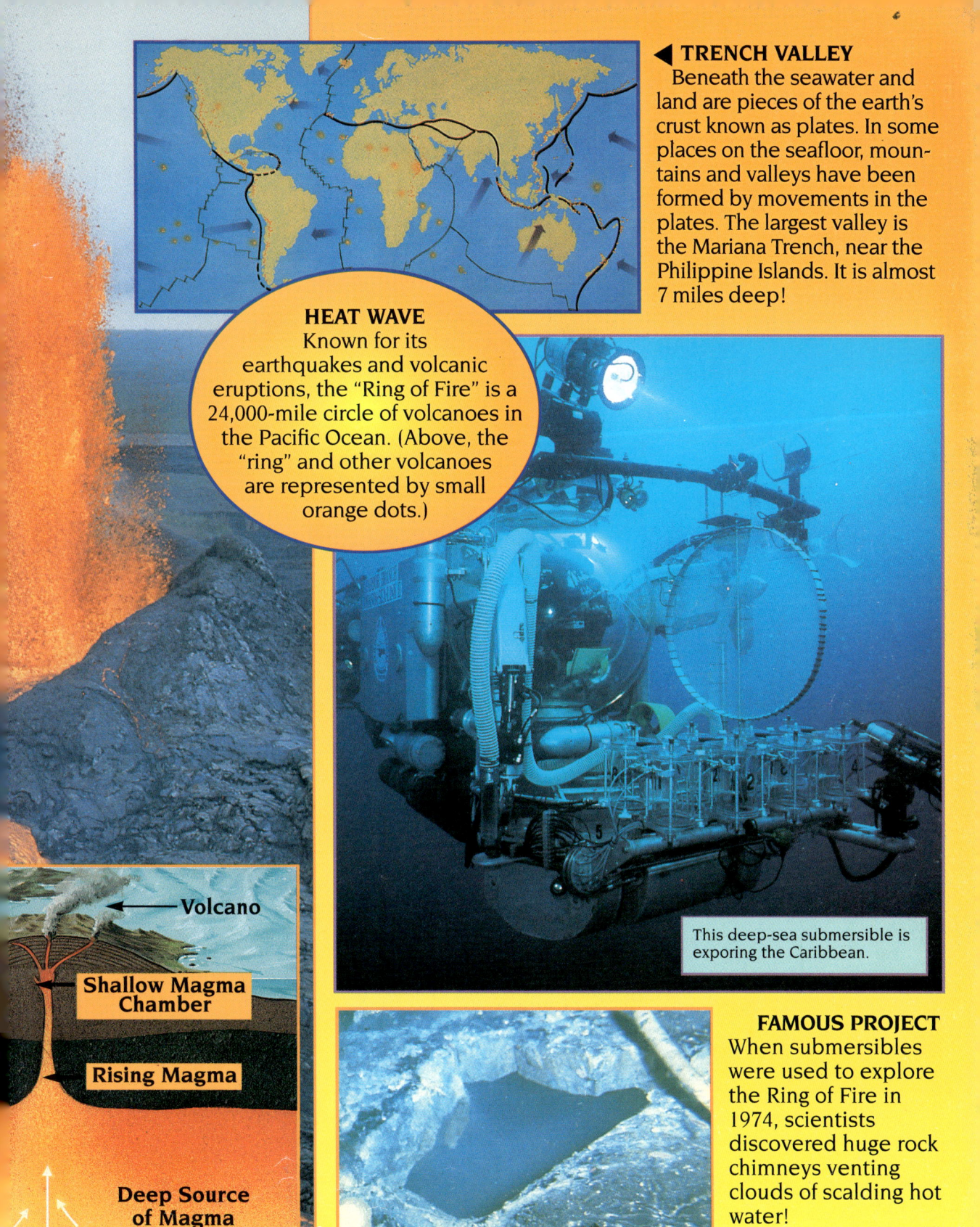

◀ **TRENCH VALLEY**
Beneath the seawater and land are pieces of the earth's crust known as plates. In some places on the seafloor, mountains and valleys have been formed by movements in the plates. The largest valley is the Mariana Trench, near the Philippine Islands. It is almost 7 miles deep!

HEAT WAVE
Known for its earthquakes and volcanic eruptions, the "Ring of Fire" is a 24,000-mile circle of volcanoes in the Pacific Ocean. (Above, the "ring" and other volcanoes are represented by small orange dots.)

- Volcano
- Shallow Magma Chamber
- Rising Magma
- Deep Source of Magma

This deep-sea submersible is exporing the Caribbean.

FAMOUS PROJECT
When submersibles were used to explore the Ring of Fire in 1974, scientists discovered huge rock chimneys venting clouds of scalding hot water!

19

TO THE BOTTOM

In previous centuries, one danger to seagoing merchants were pirates, who raided ships the world over. A great deal of loot was lost to the sea bottom. No one could dive into high-pressured, deep waters to retrieve the treasures. Today, the right equipment can get an explorer almost anywhere.

About 170 feet below the surface of the sea, scientists explored a 3,400-year-old shipwreck. Among the finds: pottery, bronze weapons, and gold. ▶

HARD HATS

In 1819 Augustus Siebe invented a copper diving helmet (weighing 20 pounds) that allowed divers to reach depths of 200 feet. A long hose, which stretched from the helmet to a pump on the surface, brought air to the diver.

AQUA LUNG

Scuba divers wear tanks containing air, which is fed into the diver's mouthpiece. This breathing device, called the *Aqua Lung*, was developed by Jacques Cousteau, the famous French oceanographer, and Emile Gagnan in 1943. It allows divers to explore the sea as deep as 500 feet.

▲ BRAVE MR. BEEBE

American explorer Charles Beebe was the first to descend to depths no diver could reach. He designed a spherical steel vessel called a bathyscape that could be lowered from a ship by a long cable. In 1934 he reached a depth of 3,028 feet!

DEEP BREATHS
In the deep sea, explorers must wear jointed metal suits that are heavy enough to withstand great water pressure. This kind of atmospheric suit makes it possible for a diver to walk on the sea bottom 2,000 feet below the surface. At right, the submersible *Star* II and the atmospheric diving suit known as JIM take divers to the bottom.

SEA LAB
Because underwater living would provide a unique view of the sea, a number of scientists have become *aquanauts*, forsaking land for a few weeks to be with the fishes. With the help of atmospheric suits, vehicles, and robots, the sea may become known in the future as the marine biologist's laboratory.

RECORD DEPTHS
Swiss physicist Auguste Piccard modified the bathyscape so that it could go even deeper. His son Jacques set the record in 1960, when he explored the Mariana Trench, 35,800 feet, almost 7 miles under the sea.

ROBOTIC EXPLORERS
Today, robot submersibles, which can descend 12,000 feet, are used to salvage treasure and explore the sea. In 1985 the *Argo* located the shipwrecked remains of the great ocean liner the *Titanic* (above), which sank in 1912. Its brother robot, *Jason*, was used to explore the wreck.

MYSTERIES OF THE DEEP

In 1872, the *Challenger* expedition of British oceanographers proved beyond a doubt that there was life deep down in the ocean. Using scoops and dredges attached to ropes, they gathered samples of 4,417 new marine organisms! But scientists have only recently begun to fathom the mysteries of the deep.

◀ This hatchetfish is being pursued by the viperfish (lower right), another deep-sea creature with light organs.

◀ The gulper fish has a huge mouth for swallowing large prey.

CLIFF HANGER
At a certain distance from each continent, the ocean floor drops sharply to a depth of 20,000 feet. In the very deep sea there is no sunlight, no plants, and the water is icy cold. Below a depth of 7,000 feet in any ocean, the temperature never rises above 39°F!

VOLCANIC CREATURES
Deep down on the ocean floor are vents that spew out scalding hot water. Warmed by liquid rock inside the earth, these springs are rich in minerals. Giant clams, tube worms twelve feet long, and blind crabs and shrimp the size of small dogs, all live near hot-water vents. They eat a special bacteria that manufactures its own food from the vent's gases and heat.

Giant tube worms

◀ NIGHT VISION
Fish in the deep sea are specially adapted to the darkness in which they live. The anglerfish, which glows in the dark, has its own "rod and bait." On the rod are lights that attract prey. A second set of teeth in the back of its throat prevent prey from escaping.

DO NOT DISTURB ▶
In 1938, fishermen in the Indian Ocean netted a coelacanth, a fish believed to have been extinct for 100 million years! Scientists speculated that it had been living undisturbed in the deep sea.

UNDER PRESSURE
Animals living in the deep ocean have adapted to the tremendous pressure of the water. Most are so perfectly adapted to this environment that they cannot survive for long when brought up to the surface—the change in pressure is just too much.

BON APPETIT ▼
Did you ever wonder who keeps the ocean floor clean? Sea cucumbers help by eating the muddy surface and digest what little food it contains.

POLAR WATERS

Have you ever been on top of the world? How about the bottom? They are pretty cold places. The North and South poles, or the Arctic and Antarctica, are known for their icebergs, glaciers, extreme weather, and polar wildlife. They also have a reputation for being dangerous to explorers!

BIG FREEZE
At any one time, there are about 200,000 icebergs floating in the Antarctic Ocean. They look big up top, but 90 percent of their volume is actually found beneath the ocean's surface. Scientists think all this water could be put to good use—to irrigate drought-stricken land, for example. But there's one problem. How do you move an iceberg?

NORTHWEST PASSAGE ▲
For hundreds of years, explorers tried to find a route along the northern coast of Canada which would link the Atlantic to the Pacific Ocean. Searching for the Northwest Passage was dangerous. Ships became frozen into the ice, or were wrecked on icebergs. A route was finally discovered in the 1850s.

This icefish can survive in waters colder than 32°F.

KEEPING WARM
What would you do to keep warm if you lived in a polar sea? Icefish, and some other polar animals, have special chemicals in their blood that prevent their body's fluids from freezing. Mammals such as whales, seals, and walruses have layers of fat, called blubber, to insulate them from harsh temperatures.

The beluga whale lives in shallow Arctic coastal waters. ▶

A KNACK FOR KAYAK
The Inuits, or Eskimos, developed the kayak—a variation on the canoe, constructed from driftwood and stretched sealskins—to hunt walrus and other Arctic animals. Today's kayaks are fiberglass, and they are paddled all over the world, not just in the Arctic!

POLAR POPULATION
Penguins live in the Antarctic. Polar bears and walruses live in the Arctic. Seals, orcas, and most whales live at both poles. In the winter, seals and walruses hunt under the ice. Seals make holes in the ice where they come up for air.

BOUNTY OF THE SEA

Treasures in the sea are not limited to those lost in shipwrecks. The ocean is a source for food, energy, and valuable minerals.

◀ Salt has been mined from the sea since ancient times.

HOMEMADE JEWELS
What happens when a grain of sand finds its way into a mollusk, such as an oyster or a conch? In one mollusk out of ten thousand, the sand becomes coated with layers of *nacre*, the stuff that forms the shell's shiny insides. Several years later a pearl is formed! To improve the odds of getting more perfect pearls, people culture the gems themselves, implanting a "seed" in an oyster around which a pearl may grow.

OIL BELOW ▶
Offshore oil wells supply about 17 percent of the world's petroleum. Most rigs are in shallow water, but deep-sea drilling techniques are being developed that could double or triple world production.

HYDRO HEAT
Looking for new energy sources, scientists are developing a new process using seawater. The heat absorbed from the sun could be stored in the water and converted into electricity.

▲ WHAT A CATCH!
A modern fisher's catch consists of fishes that swim in schools, such as tuna, salmon, anchovies, and sardines; fishes that keep to the sea floor, such as cod, haddock, and flounder; and shellfish that are harvested from shallow waters, such as oysters, clams, scallops, and lobsters.

WHALE WATCHING

In the 19th century, more than 70,000 Americans worked in the whaling industry. Whales were hunted to near extinction for their oil and other parts. Whale watching, a new popular million-dollar industry, encourages people to think ecologically.

Fishers break holes in the ice to hook passing fish.

GONE FISHING

In prehistoric times, fishers used sharpened bones for hooks, and vines for fishing line. Modern fishing fleets are like floating factories. Huge trawlers can haul in tons of fishes with each giant net. The catch is then cleaned and quick-frozen at sea.

In Mexico, these fishermen cast a net for their catch.

ENDANGERED OCEAN

All the waters of the world are connected. If one sea, or even one link in the food chain is damaged, people will also be affected. For that reason, pollution could be the real-life sea monster of the modern age.

Sewage dumped along coastlines can produce deadly bacteria in our drinking water and seafood.

DISASTER!
Oil and water don't mix, so an oil spill can spread and cover hundreds of square miles. When the *Exxon Valdez* tanker ran aground in Alaska in 1989, dumping 10 million gallons of oil, it killed hundreds of thousands of marine animals.

▶ Workers attempt to clean up the oil spilled by a tanker.

POISONED WATER
Rain washes pesticides and fertilizers from farmers' fields and homeowners' lawns into the sea, causing algae to grow out of control and destroy other living things, such as turtles and shellfish. Poisonous emissions from factory chimneys travel through the air and fall upon the land or sea as "acid rain."

CLEAN IT UP!
What's to be done about pollution in the ocean? Enforcing laws will help curb carelessness. The Clean Water Act of 1977 mandates controls and clean-ups for industrial and municipal pollution. Federal Safety Regulations, imposed by the U.S. Congress after the *Exxon Valdez* oil spill, assign the cost of clean-up to oil companies.

FISH FARM

Due to advances in technology, the ocean is in danger of being overfished. The solution may be *aquaculture*—fish farming. After they hatch from eggs, the fish are fattened up in pens until "harvest time."

The unintentional killing of other sea creatures during shrimping adds to the problem of overfishing the ocean.

A modern fish farm

SAVE THE DOLPHINS!

The fishing nets used to trap tuna often catch dolphins and other marine life, such as the shark shown above. Concern over the many dolphins killed caused major tuna canneries to stop buying tuna caught in nets. Ecologically-minded fishers use hooks instead.

MAN OF THE SEA

By exploring, scientists are learning more about the sea and the marine life it supports. One of the greatest ocean explorers was Jacques Cousteau (1910-1997). He made award-winning documentaries and wrote many books about his discoveries so that others could experience the beauty of this vast underwater realm.